The Great Seal Rescue with Lily and Zora

A Science Story About Ocean Conservation

The Great Seal Rescue with Lily and Zora

A Science Story About Ocean Conservation

A real-life inspired ocean adventure about teamwork, family, and protecting wildlife.

By Rudo Kashiri

Curious Minds Explorers Publishing

The Great Seal Rescue with Lily and Zora

A Science Story About Ocean Conservation

Illustrations by Roshanay

Published by Curious Minds Explorers Publishing

Stories That Launch Curiosity

www.curiousmindsexplorerspublishing.com

ISBN: 979-8-9996203-6-1

Printed in the United States of America

This book is a work of fiction inspired by real science. Names, characters, places, and events are used to encourage curiosity and learning and are presented in a fictionalized format.

This book is dedicated to G. Pa and G. Jeanie,

Your enduring love for marine mammals and the ocean continues to inspire Lily, Zora, and every young rescuer whose curiosity and compassion ripple outward into the world.

Thank you for demonstrating how even the smallest acts of care can create lasting waves of hope.

"When I looked out the window, I saw this beautiful planet that we live on, and I thought, 'We have to take care of this.'"
— Leland Melvin, astronaut and former NFL player

This Book Belongs To:

____

Junior Rescuer

Welcome to the Curious Minds Explorers family.
Every time you read, imagine, and learn, you help protect our oceans and the animals who call them home.

When Lily and Zora visit their grandparents, G. Jeanie and G. Pa, who rescue marine mammals, they discover how science, teamwork, and kindness can help protect ocean animals.

Back at home, their grandma Ambuya shares fun science facts and helps the girls explore the answers to the questions that spark their curiosity.

It was a rainy morning, perfect for snuggling on the couch. Ambuya sat with her tablet, and Lily and Zora cuddled close, one on each side.

"Look, girls," Ambuya said, tapping the screen. A video popped up, showing helpers gently caring for a sleepy seal pup.

Zora's eyes grew wide. "They saved it!"

Lily leaned in. "Can we help animals too, Ambuya?"

"Why do seals need rescuing?" the girls asked together.

"Sometimes ocean animals end up on the beach because they are sick or hurt," Ambuya explained. "They can get sick from germs, tangled in fishing line or trash, or from harmful algal blooms. These blooms happen when too much algae grows in the water and makes animals sick. Warmer water and pollution can make this happen more often."

"Oh no!" said Zora. "Can we help protect them?"

Ambuya smiled. "G. Jeanie just called. There's a rescue today!"

"Can we go, Ambuya?" the girls shouted.

Brody, their fluffy dog, sniffed at the tablet and wagged his tail.

"Brody wants to help, too!" Lily giggled.

"I know he does," Ambuya said gently. "But Brody can't come. We keep pets away because they might scare the animals."

RESCUE

Later that morning, Lily and Zora climbed into the car with Ambuya. The sun warmed the day, and the road hummed beneath the wheels.

"Are we really going to help a seal?" Lily asked, bouncing in her seat.

Ambuya nodded. "Yes, my little explorers. Today, you'll learn what it means to be a Junior Rescuer."

Zora hugged her backpack. "I'm ready!"

When they arrived at the beach house, G. Jeanie and G. Pa were waiting outside with warm hugs.

"Welcome, Junior Rescuers!" G. Pa said proudly.

Just then, a black cat tiptoed out.

"Hello, Chairman Meow," Lily giggled.

Zora laughed. "Are you ready to help too?"

G. Jeanie smiled. "We got a call about a stranded seal pup. Are you ready to go see how we can help?"

Lily and Zora looked at each other and took a deep breath. "We're ready," they said.

"G. Jeanie and I are trained rescuers," said G. Pa. "We'll show you what to do."

G. Jeanie clapped her hands. "First, we need to learn about the tools," she said.

Marine Rescue
PROTECT
OUR OCEANS

G. Jeanie opened the small shed door. "Come on in," she said.

Inside were gloves, cones, towels, a crate, and rescue charts.

Zora pointed at the crate. "Is that for the seals?"

"Yes," said G. Pa. "It helps us carry them safely without touching them."

Lily picked up a bright orange cone. "What are these for?"

"To make a safe space," G. Jeanie said. "They remind people to stay back."

"We're going to help an elephant seal pup," G. Pa said. "Elephant seal mothers nurse their pups for about five weeks. They stay together on the beach in a rookery, kind of like a big seal neighborhood."

"One day, the mother heads back to the ocean, and the pup is on its own. So we give them lots of space while the mother is still there. If a pup needs help, trained rescuers step in, usually after it is weaned."

Lily tilted her head. "So if Mom leaves, how does the pup learn to swim?"

G. Pa smiled. "Good question. It learns step by step. First, it practices in shallow water near shore. Then it grows stronger and learns to catch fish."

Zora gasped. "And how can you tell when a pup really needs help?"

"We watch from far away. If the pup looks hurt, tangled, or alone for a long time, then it may need help, so we call our rescue team," G. Pa said.

At the beach, the rescue truck rolled in. Doors clicked open softly. Rescuers stepped out with bright blue and yellow rescue boards and a padded transport crate, moving slowly and quietly across the sand.

Zora pointed. "They're making a safety zone!"

"So no one gets too close," Lily whispered.

G. Jeanie nodded. "That's right. Even when we want to help, we have to respect wild animals and give them space."

Right away, the rescue team set the boards down to create a safe path, like a gentle hallway through the sand. They spoke in calm, steady voices so the seal pup would not startle or run away.

Step by step, they guided the pup closer to the crate.

Lily squeezed Zora's hand. "Do you think it's scared?"

"Maybe a little," Zora whispered. "But look how careful they are."

From far away, one rescuer gave the twins a quick thumbs-up.

G. Pa gave them a proud wink. "Good teamwork, girls," he whispered. "Watching quietly helps the rescuers do their job."

The rescue team gently loaded the pup's crate into a truck.

G. Jeanie walked the girls to their car. Lily and Zora looked back at the beach, proud and quiet.

"Let's follow the rescue truck to the Marine Mammal Hospital," G. Pa said as everyone buckled up.

The family's car trailed behind the truck, winding along the sparkling coastline.

Zora pressed her face to the window. "Do you think it's okay?" she whispered.

"Maybe," Lily said softly. "But it's not alone. The rescuers will take care of it."

As the road curved, the hospital came into view. Lily's eyes widened. "Do they help other animals here too?" she asked.

G. Pa nodded. "They do. They help lots of marine mammals, like harbor seals and sea lions."

"Sometimes a harbor seal pup rests on the sand while its mom swims out to find food. Most of the time, its mom comes back, so we give the little seal lots of space."

"But if people or dogs get too close again and again, its mom might stay far away or not come back."

Zora blinked. "What if its mom doesn't come back?"

"First, we wait and watch from far away," G. Pa said. "Sometimes its mom is still nearby and comes back. But if the pup is alone for a long time, hurt, or tangled, that's when we call trained rescuers."

The Marine Mammal
Hospital
MARINE MAMMAL RESCUE

Soon, the cars pulled into the hospital parking lot. The rescue team carried the seal's crate into the bright, cheerful hospital. Colorful posters and charts covered the walls, showing different ways people can help animals.

Lily and Zora hurried after them, excited and a tiny bit nervous.

Inside, animal doctors and helpers smiled and waved at the twins.

"Welcome, little seal," said Dr. Tandi softly.

She turned to Lily and Zora. "Would you like to help us choose a name for this new friend?"

Lily's eyes lit up. She leaned close to Zora and said, "Let's call it Luna."

Then Lily looked up at Dr. Tandi. "Luna means moon," she explained. "The moon helps you find your way, and we're helping Luna find the way home."

Dr. Tandi smiled. "Luna is a wonderful name."

Zora watched a playful seal splash in a nearby pool. Then she looked back at the crate. "Will Luna get better?" she asked.

Dr. Tandi nodded. "I'm glad you asked," she said. "Luna is in the right place. We're going to help her rest, eat, and get strong. And when she's ready, we'll take her back home to the ocean."

Lily squeezed Zora's hand. They looked at each other and smiled.

The Marine Mammal Center
SAVE THE OCEAN
HOW YOU CAN HELP THE OCEAN
Clean the Beach
Use Fewer Plastic Products
Clean the Beach
Save Water
Save Sea lions

Zora looked around, then turned to Dr. Tandi. "What are you going to do to take care of Luna?"

"First, we find out what's wrong," Dr. Tandi said, pointing to a chart of a seal's body. "Seals can get sick in different ways."

Lily traced the picture with her finger. "And sometimes a pup isn't sick," she said, remembering what Ambuya taught them. "Sometimes it's still learning to swim and find food. If it gets too skinny, dehydrated, or can't find enough fish to eat, the rescue team steps in."

Dr. Tandi nodded. "That's right. Seals don't drink seawater. They get the water they need from the fish they eat."

Zora nodded. "So food gives them the water they need and keeps them from getting dehydrated."

Dr. Tandi clapped softly. "Junior Rescuers, time for a quick challenge!" she said. "When I say, 'Seal on the sand!' you say what?"

Lily stood up straighter. "Tell a grown-up!" she called.

Zora raised her voice too. "And no touching. *Give the seal space!*"

Dr. Tandi beamed. "Perfect. *Give seals space and let a grown-up call the rescue team.*"

Lily and Zora smiled. They were happy they got the challenge right.

Lily and Zora walked quietly past pools where seals splashed and dove.

Zora leaned closer to Dr. Tandi. "What else do you do to help the seals?"

Dr. Tandi led them to the pools and pointed. "We give them safe enrichment toys," she said. "Some look like pretend seaweed. Some are floating balls with a little fish tucked inside."

Lily pointed to a bright, colorful ball bobbing on the water. "Like that one."

A pup nudged the ball with its nose. Plop! A tiny fish popped out, and the pup gulped it up.

Zora leaned in and laughed. "It looks like they're playing ball!"

Dr. Tandi nodded. "Yes. The toys keep their minds and bodies busy while they heal. And if a seal is sick or weak, this is a safe place to rest and get strong before heading back to the ocean."

Lily nodded. "So this is their learning pool."

Dr. Tandi smiled. "Exactly," she said. "Learn here first, then back to the wild."

Zora's eyes sparkled. "I hope Luna gets to swim again soon."

"She will," Dr. Tandi said. "Luna just needs time, care, and teamwork."

Lily and Zora smiled at Luna's crate and then at the splashing pups. They couldn't wait to see her grow stronger.

The hospital buzzed with splashes and seal barks. Lily and Zora watched Luna wiggle and sniff the air.

"I think she's hungry," Zora said.

Lily nodded. "And maybe she misses her mom."

Dr. Tandi walked over carrying a shiny bucket brimming with fish. "It's feeding time," she said. "Would you like to help?"

She handed Lily and Zora a scoop each.

"We wear face coverings," Dr. Tandi said, "so the seals do not learn to expect food from people."

The girls counted together: one, two, three, giggling as they dropped fish into their scoops.

Splash! A playful pup popped up and caught a fish midair, sending sparkles of water everywhere.

"Did you see that?" Lily and Zora squealed together.

The seals slapped their flippers. One balanced a fish on its nose. Another spun in a silly circle.

"Let's cheer for the seals!" Dr. Tandi said.

Lily and Zora cheered and made funny "Arrf! Arrf!" sounds.

Dr. Tandi smiled. "For now, they eat lots of fish and grow stronger every day."

A few weeks later, G. Jeanie called with happy news. "Luna is healthy, and she's ready for release day!"

Lily and Zora smiled at each other and whispered, "Yay, Luna."

Back at the beach, everyone formed a quiet line near the shoreline as Luna's crate was carried toward the water. The waves whooshed and foamed, and everyone held their breath.

Zora held Lily's hand. "This feels really important," she whispered.

Lily nodded. "We have to stay still, just like before."

G. Jeanie stood beside the twins. "Are you ready?" she asked softly.

The girls squeezed each other's hands and took a slow, deep breath.

The crate door swung open. For a moment, Luna peeked out and blinked at the bright sunlight and sparkling waves. Then she stepped onto the sand, waddle, waddle, and headed for the ocean.

"Goodbye, Luna," Lily whispered. "Be brave out there."

Zora's voice turned soft. "Swim home, Luna," she said. "We're proud of you."

Luna splashed into the water and swam forward, stronger now.

Lily wrapped her arms around Zora. "She's home," she whispered.

Together, they watched until Luna was just a little dot in the waves.

Soon the sun dipped down, and the beach grew quiet again. G. Jeanie sat with Lily and Zora on a big driftwood log. The ocean shone softly, calm and smooth. Far away, a seal rested on the rocks.

G. Jeanie hugged Lily and Zora close. "Girls, what did you learn from this adventure?" she asked.

Zora smiled. "We should remind people not to feed marine mammals, because human food can make them sick."

Lily added, "And wild animals need space, so we watch quietly and never get too close."

G. Jeanie smiled. "That's right. We also learned that seal pups need their moms, and when animals need help, rescue teams know what to do."

"Even small helpers like you can make a big difference," she said. "I'm so proud of you both."

Lily and Zora hugged G. Jeanie and said together, "Thank you for teaching us."

The twins looked at each other and smiled, proud to be Junior Rescuers. They knew their voices, their choices, and their kindness could help protect seals like Luna, today and every day.

That night, back at the beach house, the kitchen felt warm and cozy after G. Pa's super-yummy lasagna. Everyone gathered around the table. A big, crinkly coastline map lay open like a treasure chart.

Ambuya smiled. "I'm proud of my ocean explorers," she said softly. "When we help one animal, we help the ocean."

G. Pa nodded. "You asked a lot of questions, and you learned a lot about what it takes to be a rescuer," he said. "And you helped, too."

Lily looked at Zora and whispered, "I want to help ocean animals forever."

Zora grinned. "We're Junior Rescuers," she said.

The twins hugged their grandparents tight. They were happy that Luna was safe and back home in the ocean.

Brody opened his mouth wide. "Aaaawoooo," he yawned, then curled up under the table.

Everyone laughed softly.

For a moment, everyone just breathed and listened to the ocean outside.

Ambuya tapped the map. "Tonight, we can help in a new way," she said. "Let's make posters to remind everyone what we learned."

"Yes, Ambuya!" they both said. "Let's make them tonight, before we forget any of our ideas."

What would you put on your own poster to help protect the ocean or animals?

That night, Lily and Zora spread out their art supplies on the table.

"Crayons!" Lily cheered as markers rolled across the table. Seashells sat beside scraps of recycled paper, ready to be glued on.

"We're making Ocean Protector posters!" Lily announced.

Lily wrote in big letters:

Protect Our Ocean!

Don't Feed Sea Animals!

Zora added her own signs:

No Plastic on the Beach!

Be Kind to Sea Creatures!

"We can tell our friends to pick up litter," Lily said, carefully gluing a shell to her poster.

"And use reusable items, with no plastic left behind," Zora said, drawing a giant blue wave.

Just then, Brody trotted over, *tap, tap, tap*, and left a sandy paw print on the edge of a poster.

"Brody signed it!" the girls laughed. "Perfect!"

They taped their posters up in the window, proud and smiling.

"We're Junior Rescuers at home and at the beach," Zora said.

Lily nodded. "Every day."

Save Our Ocean!
Keep the Beach Clean
No plastic on the Beach
Be Kind to Sea Creatures
Protect Our Ocean
Don't Feed Sea Animals

G. Pa stood beside the twins. "What you've learned matters," he said. "Your voices, your choices, and your kindness can help protect wild animals."

He turned to the twins and smiled proudly. "Your posters share an important message," G. Pa said. "Now let's put that message into action. Choose one Ocean Protector step today, and remember that small helpers can make a big difference."

Lily and Zora smiled back. Hand in hand, they said, "We're ready to help."

G. Pa's Beach Safety Tips

At the beach, you're visiting an ocean animal's home. Here's how to keep **them** safe, and keep **you** safe, too.

□ **TIP 1: Give Seals Space**
Seals are wild animals. Getting too close can scare them and cause stress.
G. Pa says: *"Always stay at least* 50 yards (*150 feet) away."*

□ **TIP 2: Never Touch Seal Pups**
A seal pup may look lonely, but its mother is often nearby hunting for food. Getting too close can scare the mother and make it harder for her to return.
G. Pa says: *"Looking from far away helps pups and their moms stay together."*

□ **TIP 3: Call for Help**
If you see a seal that looks sick, hurt, or stranded, do not try to help by yourself.
G. Pa says: *"Ask a grown-up to call a wildlife rescue center. Trained helpers know what to do."*

□ **TIP 4: Never Feed Wild Animals**
Don't share food with wild animals. Human snacks can make them sick, and it can teach them to swim up to people for food.
G. Pa says: *"A seal snack is fish... not french fries!"*

□ **TIP 5: Leave No Trace**
Pick up your snacks, wrappers, and strings so animals don't swallow them.
G. Pa says: *"If you bring it in, take it out!"*

□ **TIP 6: Keep Dogs Away**
Ocean animals can get scared if dogs run up to them.
G. Pa says: *"Keep pets on a leash and far away from wildlife."*

Junior Rescuer Pledge

I promise to protect our blue planet.

I will keep beaches clean and help wildlife stay safe.

I will share what I learn and teach others to care for the ocean.

With curiosity, kindness, and teamwork,

I can make a difference, one small action at a time!

Your Ocean Promise:

✏ Write (or draw!) your own promise here:

Try It! Make Your Own Ocean Rescuer Poster

Lily and Zora have a challenge for you! Make a poster that reminds everyone how to keep beaches safe for sea animals.

What you need

☐ Paper or recycled cardboard (like a cereal box)

☐ Markers, crayons, or colored pencils

☐ Optional: glue or tape

☐ Optional: clean seashells (ask a grown-up first)

Steps

1. Choose your message (pick 1 to 3 ocean rules).
2. Write big, bold words that are easy to read from far away.
3. Add a picture (waves, a sea lion, a trash bag, or a "No Feeding" sign).
4. Decorate with color, borders, and fun patterns.
5. Share your poster: hang it in a window or classroom, or take a photo.

Poster message ideas (pick one or make your own)

- Protect Our Ocean!
- Give Seals Space!
- Pack It In, Pack It Out!
- No Plastic on the Beach!
- Be Kind to Sea Creatures!
- Pick Up 5 Pieces of Trash!

Write your big message at the top, then draw pictures below. Add arrows or labels if you want.

Author's Note

Seals are playful, intelligent animals found along the California coast and in many places around the world. Sometimes they need our help, whether they are sick, tangled in fishing line, or affected by pollution.

In this story, we see how science, teamwork, and kindness can protect seals and the ocean they call home. Small actions, like picking up trash, calling for help, and sharing what you know, can make a big difference for animals and the environment.

Rescue organizations and volunteers like G. Jeanie and G. Pa work hard every day to help seals heal and return to their ocean home.

As a former NASA educator and a lifelong science storyteller, I believe curiosity and compassion go hand in hand. When children discover how science connects to caring for our planet, they grow into problem-solvers and protectors of all living things.

Remember: you can make a difference.
Learn, observe, protect, and share what you discover. Every small action helps keep our oceans healthy.

Rudo Kashiri
Curious Minds Explorers Publishing

www.ingramcontent.com/pod-product-compliance
Lightning Source LLC
LaVergne TN
LVHW070859210926
848482LV00016B/976
9798999620361